A PEEK AT PLANTS

A PEEK AT FRUITS

DOUG BRADLEY

PowerKiDS press

Published in 2026 by The Rosen Publishing Group, Inc.
2544 Clinton Street, Buffalo, NY 14224

First Edition

Editor: Greg Roza
Book Design: Leslie Taylor

Photo Credits: Cover, p. 1 Maya Roy/Shutterstock.com; p. 5 brgfx/Shutterstock.com; p. 7 (blueberries) Volosina/Shutterstock.com, (watermelon) pukao/Shutterstock.com, (tomatoes) Photoongraphy/Shutterstock.com, (cucumbers) New Africa/Shuttertock.com; p. 9 imageBROKER.com/Shutterstock.com; p. 11 Oleksiichik /Shutterstock.com, (inset) MarcoFood/Shutterstock.com; p. 13 Basia Seaman/Shutterstock.com; p. 15 Taras Vykhopen/Shutterstock.com; p. 17 Doug McLean/Shutterstock.com; p. 19 (strawberry bush) Tim UR/ Shutterstock.com, (orange trees) GUDAR'S MILLER/Shutterstock.com, (grapes on vine) digitizesc/Shutterstock.com; p. 21 digitizesc/Shutterstock.com.

Library of Congress Cataloging-in-Publication Data

Names: Bradley, Doug, 1971- author.
Title: A peek at fruits / Doug Bradley.
Description: [Buffalo] : PowerKids Press, [2026] | Series: A peek at plants | Includes bibliographical references and index.
Identifiers: LCCN 2024056072 (print) | LCCN 2024056073 (ebook) | ISBN 9781499451986 (library binding) | ISBN 9781499451979 (paperback) | ISBN 9781499451993 (ebook)
Subjects: LCSH: Fruit–Juvenile literature.
Classification: LCC QK660 .B73 2026 (print) | LCC QK660 (ebook) | DDC 575.6/7–dc23/eng/20250107
LC record available at https://lccn.loc.gov/2024056072
LC ebook record available at https://lccn.loc.gov/2024056073

Manufactured in the United States of America

CPSIA Compliance Information: Batch #CSPK26. For Further Information contact Rosen Publishing at 1-800-237-9932.

CONTENTS

Plants and Their Parts

There are so many plants in the world! Most plants have the same parts. Stems allow a plant to stand tall. Leaves make food for the plant. Roots keep the plant in the ground and take in water. Many plants also have flowers and fruits. Let's learn about fruits!

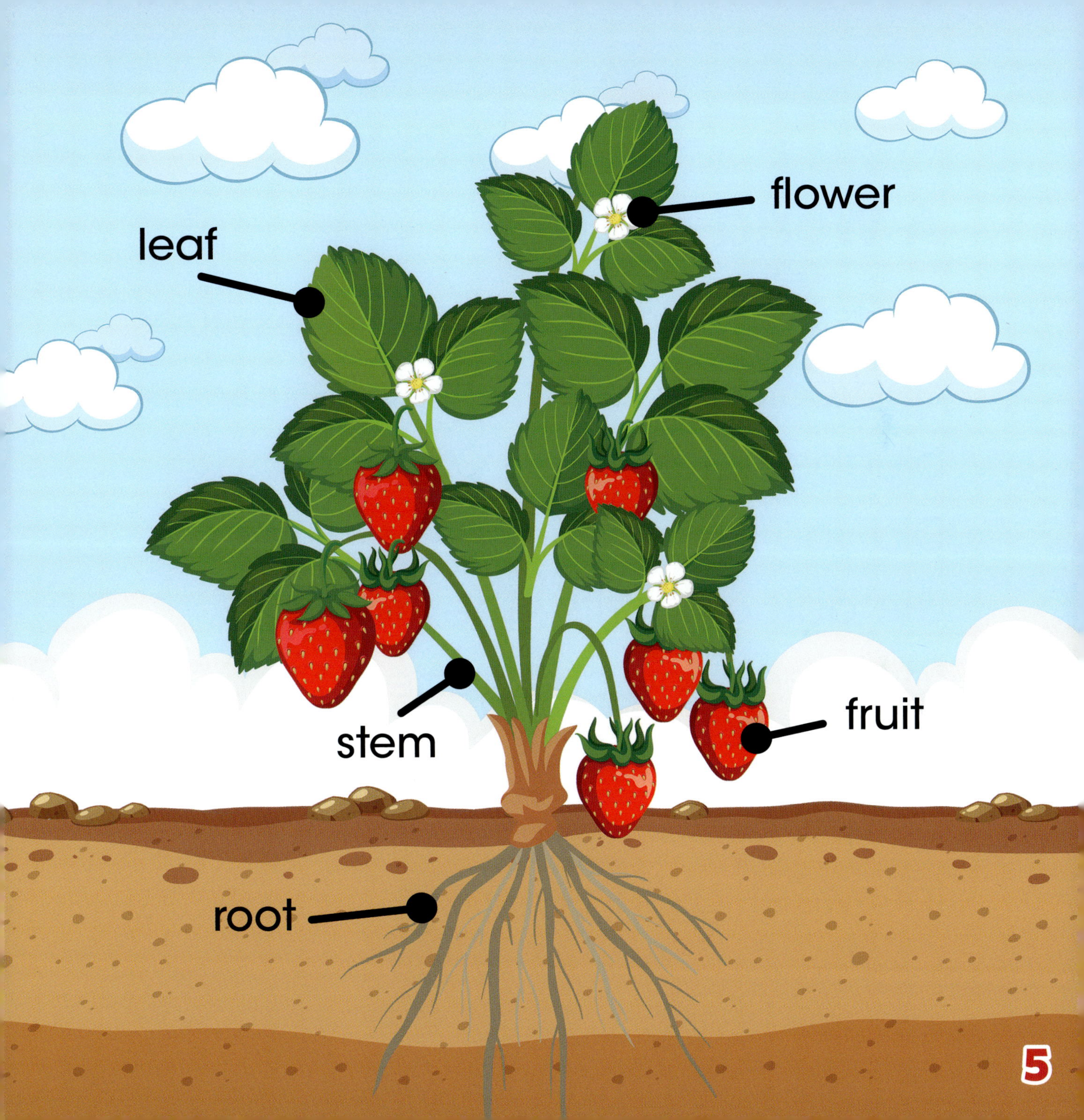
flower
leaf
stem
fruit
root

A World of Fruit

Fruits come in different shapes, sizes, and colors. You may think of tiny blueberries or large watermelons! Did you know many vegetables are actually fruits? They include tomatoes and cucumbers. Even some nuts are fruits! However, people usually use the word "fruit" for sweet-tasting plants.

From Flower to Fruit

Fruits start as flowers! Flowers have colorful leaves called petals. Inside the petals is a special part where the plant's seeds begin to form. Soon, the petals **wither** and fall away. What's left is the part of the plant where seeds grow. It's called the ovary.

pear blossoms

Once the petals are gone, the ovary keeps growing larger. The seeds grow inside the ovary. Some seeds are small. Some ovaries grow a single, large seed called a pit. Some ovaries grow many seeds. Once the ovary is finished growing, it's a fruit!

pears
pear
seeds

Safe Seeds

Fruits grow to keep seeds safe. Many fruits have a hard outer skin or rind. Some fruit-bearing plants have **thorns** to scare away animals. Some fruits just don't taste good! Some fruits, such as the winterberry, are **poisonous**. Winterberries are bright red, which tells animals to stay away!

Rotting

It can take days or weeks for fruit to become ripe. In time, the fruit starts to rot. When fruit has been picked off of a plant, it will rot soon after. Farms that grow fruit often keep them cold so they last longer.

Spreading Seeds

When rotten fruit drops to the ground, the seeds can start to grow. Some animals eat the fruit that grows on trees and **bushes**. They leave the seeds in another place when they poop. Some seeds or fruits stick to animals' fur and are dropped in a new place.

Fruit Farms

Farmers grow many kinds of fruit. Some fruits, such as berries, grow on bushes. Some fruits, such as grapes, grow on **vines**. Other fruits grow on trees. A farm with fruit trees is called an orchard. Common kinds of orchards are apple, pear, peach, cherry, and orange orchards.

strawberry bush

orange tree

grapes on vine

Grow Your Own!

You may live where fruits, such as blackberries, grow in the wild. But you can grow your own fruit! Strawberry bushes and tomato vines are common plants to grow at home. You can also plant an apple tree, but you'll need to wait several years before you can enjoy fresh apples.

GLOSSARY

bush: A low plant with many branches and closely packed leaves.

poisonous: Including poison, or a substance that can make animals sick when eaten or touched.

thorn: A sharp, pointed part of a plant, often found on its stem or branches.

vine: A bendable stem that can creep along the ground or climb up a wall.

wither: To dry up and grow smaller.

FOR MORE INFORMATION

BOOKS

Quick, J. F. *Apples*. Mankato, MN: The Creative Company, 2024.

Silvora Collections. *Plant Parts and their Purpose: A Budding Botanist's Activity Book*. Independently published, 2024.

WEBSITES

Facts for Kids: Fruits
factsforkids.com/category/fruits
Discover facts about your favorite fruits at this fun, detailed website.

Plant Life Cycle
www.activewild.com/plant-life-cycle-kids
You can read much more about how plants grow at this website.

INDEX